Air Power

EUROFIGHTER TYPHOON

MEGAN COOLEY PETERSON

BLACK RABBIT BOOKS

Bolt is published by Black Rabbit Books
P.O. Box 3263, Mankato, Minnesota, 56002.
www.blackrabbitbooks.com
Copyright © 2020 Black Rabbit Books

Jennifer Besel & Marysa Storm, editors;
Grant Gould, designer; Omay Ayres,
photo researcher

All rights reserved. No part of this book may be reproduced, stored in a retrieval system or transmitted in any form or by any means, electronic, mechanical, photocopying, recording, or otherwise, without written permission from the publisher.

Library of Congress Cataloging-in-Publication Data
Names: Peterson, Megan Cooley, author.
Title: Eurofighter Typhoon / by Megan Cooley Peterson.
Description: Mankato, Minnesota : Black Rabbit Books, [2020] | Series: Bolt. Air power | Includes bibliographical references and index. | Audience: Grades 4-6. | Audience: Ages 8-12.
Identifiers: LCCN 2018030434 (print) | LCCN 2018030746 (ebook) | ISBN 9781680727937 (e-book) | ISBN 9781680727876 (library binding) | ISBN 9781644660102 (paperback)
Subjects: LCSH: Eurofighter Typhoon (Jet fighter plane)—Juvenile literature.Classification: LCC UG1242.F5 (ebook) | LCC UG1242.F5 F4724 2020 (print) | DDC 623.74/644—dc23
LC record available at https://lccn.loc.gov/2018030434

Printed in the United States. 1/19

Image Credits

Alamy: Jason O. Watson (Aviation Photos), 28; Mark Kelly, 31; Open Government Licence, 20; Stuart Abraham, 3, 18; commons.wikimedia.org: Sugar-monster, 14 (b); commons.wikimedia.org/defenceimagery. mod.uk: RAF, 24–25; defense.info: Airbus Defence and Space, 4–5; difesa.it: Aeronautica Militare, 14 (t); eurofighter.com: Eurofighter, 15 (bl); Eurofighter Typhoon Multimedia, 6; Geoffrey Lee, Planefocus Ltd/ Eurofighter, Cover, 15 (br); Geoffrey Lee/Eurofighter, 15 (tl); Jamie Hunter /Eurofighter, 15 (tr); flickr.com: poter.simon, 22; leonardocompany.com: Leonardo/ Aerostructures Division, 26 (t); libertycity.net: SkylineGTRFreak & Oskar, 12 (jet); nae.fr, 27 (b); Newscom: Andy Rouse/Avalon, 32; MAAA/ ZDS/Wheatley/WENN, 4; Shutterstock: David Schwimbeck, 18; Elenamiv, 12 (bkgd); Gary Blakeley, 11; GERSHBERG Yuri, 22; inspired-fiona, 21 (bgkd); Matteo Chinellato, 1; Ryan Fletcher, 26–27; sam-whitfield1, 14–15 (bkgd); vaalaa, 17; the-blueprints: The Blueprints, 20–21; turbosquid.com: 3d_molier International, 8–9
Every effort has been made to contact copyright holders for material reproduced in this book. Any omissions will be rectified in subsequent printings if notice is given to the publisher.

CONTENTS

CHAPTER 1
A Sneaky Fighter Jet....4

CHAPTER 2
History of the
Eurofighter Typhoon...10

CHAPTER 3
Eurofighter Typhoon
Features................16

CHAPTER 4
The Eurofighter
Typhoon in Action.....23

Other Resources...........30

CHAPTER 1

A SNEAKY Fighter Jet

It was December 2017. British **prime minister** Theresa May boarded her airplane. The plane took off. Suddenly, two Eurofighter Typhoons pulled up beside her plane.

Showing Off

The pilots tipped the Typhoons' wings. May saw the fighters' weapons. But she wasn't afraid. She knew this was a fake mission. The Typhoon pilots were showing May how easily they could sneak up.

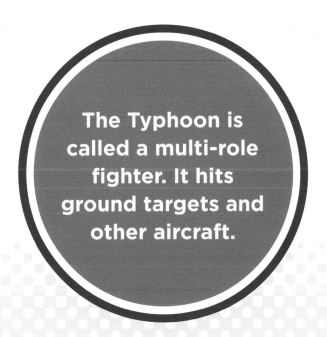

The Typhoon is called a multi-role fighter. It hits ground targets and other aircraft.

TYPHOON PARTS

WINGS

COCKPIT

FUEL TANKS

CANNON

CHAPTER 2

HISTORY of the Eurofighter Typhoon

In the 1970s, **European** fighter jets were getting old. They weren't fast enough. They couldn't carry many weapons. But the cost to build a new high-tech jet was high.

Working Together

The countries of Germany, Italy, Spain, and the United Kingdom joined forces. They worked together to make a new jet. Each country helped build part of it. They named the new jet the Eurofighter Typhoon.

**Nations that
Fly the Typhoon**
(as of 2018)

Austria
Germany
Italy
Kingdom of Saudi Arabia
Oman
Spain
United Kingdom

TYPHOON TIMELINE

1997
Pilots do the 500th test flight.

1980

1989
The first test models are built.

2003
The fighter enters active service.

2006
Typhoons fly over the Olympic Games. They keep athletes safe from possible threats.

2020

2016
Kuwait orders 28 fighters.

2017
Qatar orders 24 Typhoons.

15

CHAPTER 3

Eurofighter Typhoon FEATURES

The Typhoon zips past other aircraft. It can fly up to 1,535 miles (2,470 kilometers) per hour.

Most fighters use **afterburners** to go **supersonic**. Afterburners give a plane more **thrust**. But they use a lot of fuel. The Typhoon goes supersonic without afterburners.

COMPARING FIGHTER JET TOP SPEEDS

Eurofighter Typhoon
F-22 Raptor
F-35 Lightning II
MiG-35

top speed

1,535 miles (2,470 km) per hour

about 1,500 miles (2,414 km) per hour

1,200 miles (1,931 km) per hour

more than 1,600 miles (2,575 km) per hour

1,000   1,100   1,200   1,300   1,400   1,500   1,600

The Typhoon's Weapons

short-range air-to-air missiles · · · · · · · · ·

beyond-visual-range air-to-air missiles · · · ·

laser-guided bombs · · · · · · ·

27mm Mauser cannon

Weapons

The Typhoon is loaded with weapons. Missiles shoot down enemy planes. Lasers guide missiles to targets. The fighters also drop bombs on ground targets.

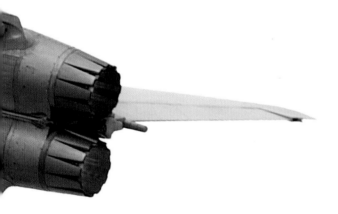

The fighter can carry more than 16,000 pounds (7,257 kilograms) of weapons.

Sensors and Radar

The Typhoon uses high-tech equipment. **Sensors** and **radar** warn pilots of incoming missiles. They also spot enemies. The Typhoon can attack before being seen.

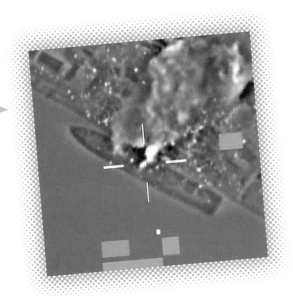

length
52.4 feet (16 m)

TYPHOON STATS

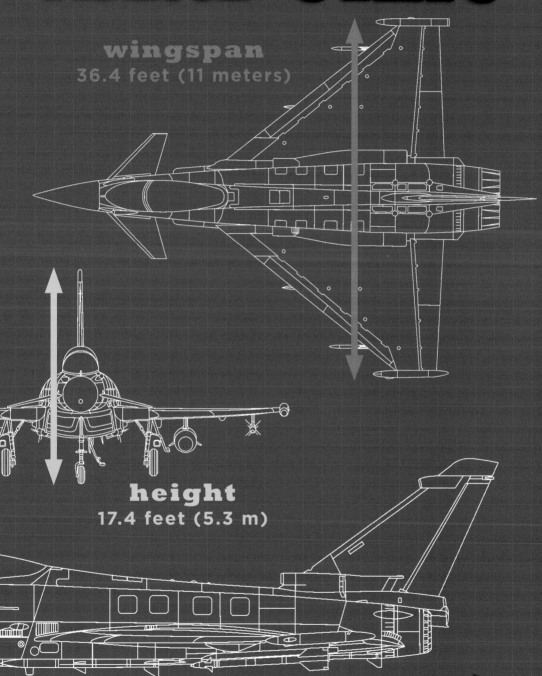

wingspan
36.4 feet (11 meters)

height
17.4 feet (5.3 m)

CHAPTER 4

The Eurofighter Typhoon in ACTION

Typhoons flew their first missions in 2011. They bombed **terrorists** in Libya. Typhoons flew more than 600 missions in six months of the operation.

Scaring Enemies Away

In 2017, an unidentified aircraft flew over the Baltic. A Typhoon was called to help. Within seconds, the jet reached 40,000 feet (12,192 m). It pulled up by the aircraft, scaring it away.

The Baltic includes the countries Estonia, Latvia, and Lithuania.

BY THE NUMBERS

400 — NUMBER OF COMPANIES THAT BUILD PARTS FOR THE FIGHTER

55,000 feet (16,764 m) — MAXIMUM FLYING HEIGHT

about **$140 MILLION** cost to buy one Typhoon

11 NUMBER OF DAYS IT TAKES TO PAINT EACH FIGHTER

Powerful Fighters

Typhoons are advanced fighter jets. They win **dogfights** and attack targets in a single mission. Typhoons work hard to keep countries safe.

GLOSSARY

afterburner (AF-tur-burn-uhr)—a device in the tailpipe of a turbojet engine that injects fuel into the hot gases to provide extra thrust

dogfight (DAWG-fiyt)—a mid-air battle between fighter planes

European (yur-oh-PEE-uhn)—relating to the continent of Europe or its people

prime minister (PRIYM MIN-uh-ster)—an official head of a government cabinet or ministry

radar (RAY-dar)—a device that sends out radio waves for finding the location and speed of a moving object

sensor (SEN-sor)—a device that finds heat, light, sound, motion, or other things

supersonic (soo-pur-SON-ik)—faster than the speed of sound; the speed of sound is about 761 miles (1,225 km) per hour.

terrorist (TER-ur-ist)—a person who uses violent acts to frighten people in order to achieve a goal

thrust (THRUHST)—the force that pushes a vehicle forward

LEARN MORE

BOOKS

Harris, Tim. *Superfast Jets: From Fighter Jets to Turbojets.* Feats of Flight. Minneapolis: Hungry Tomato, 2018.

Peterson, Megan Cooley. *F-22 Raptor.* Air Power. Mankato, MN: Black Rabbit Books, 2019.

Willis, John. *Fighter Jets.* Mighty Military Machines. New York: AV2 by Weigl, 2017.

WEBSITES

Eurofighter Typhoon
www.baesystems.com/en/product/typhoon

Eurofighter Typhoon
www.eurofighter.com/

Typhoon FGR4
www.raf.mod.uk/aircraft/typhoon-fgr4/

INDEX

C

costs, 10, 27

F

features, 8–9, 20

H

history, 10, 13, 14–15

M

missions, 4, 7, 15, 23, 24

S

sizes, 20–21

speeds, 16–17

W

weapons, 7, 8–9, 18, 19